~~About the Author ~~

I have five brothers and five sisters.
How many children do my parents have altogether?

(a)   10   (b)   12   (c)   11   (d)   9

I am 41 years old and my youngest sibling is 9 years younger.
How old is my youngest sibling?

$$41$$
$$-\ \underline{\quad 9 \quad}$$

Deadria, Devin and their Dad went to church; they picked up 3 of Jesus' children along the way.
How many people went to church together?

(a)   6   (b)   7   (c)   5   (d)   4

~~Special Trivia~~

My middle name is <u>D</u>eloris; I wish it was <u>D</u>allas though! My daughter is <u>D</u>eadria and my son is <u>D</u>evin.
What do you think is my favorite letter of the alphabet? Circle the letter you think it is.

(1) A   (2) C   (3) M   (4) D

What is your favorite letter of the alphabet? _____

Write a sentence using your favorite letter of the alphabet-

~~Special Thanks~~

First, I want to give thanks to the Lord for giving me the idea, strength and courage to pursue this incredible journey. Without him I could not have accomplished this goal. Thank you Lord!

Next, I would like to thank my wonderful husband Michael and my children Deadria and Devin who supported me throughout this process. And finally, I would like to thank my friends for being understanding. Thank you all for your love and support!

~~Dedication~~

This book is especially dedicated to my son Devin who is 8 years old and loves mathematics. I want Devin to grow up with a spiritual foundation so that he is grounded in the word of God. My hope is that he will have a balance life, where he can apply the biblical teachings to his everyday life. What better way to accomplish this than taking the scholastic subject he loves best and relate it to the words of the Lord? As a child, I felt a true connection with God and used the biblical teachings to stay focused and to get to know God on a more personal level. I want my son to have those special memories as well.

Love,
VDG

~~Books of the Bible~~

The Bible is broken down into the Old and New Testament. There are **66** Books in the Bible, There are **39** in the Old and ___ in the New. How many books are in the New Testament?

Old Testament

1. Genesis
2. Exodus
3. Leviticus
4. Numbers
5. Deuteronomy
6. Joshua
7. Judges
8. Ruth
9. 1 Samuel
10. 2 Samuel
11. 1 Kings
12. 2 Kings
13. 1 Chronicles
14. 2 Chronicles
15. Ezra
16. Nehemiah
17. Esther
18. Job
19. Psalm
20. Proverbs
21. Ecclesiastes
22. Song of Solomon
23. Isaiah
24. Jeremiah
25. Lamentations
26. Ezekiel
27. Daniel
28. Hosea
29. Joel
30. Amos
31. Obadiah
32. Jonah
33. Micah
34. Nahum
35. Habakkuk
36. Zephaniah
37. Haggai
38. Zechariah
39. Malachi

New Testament

1. Matthew
2. Mark
3. Luke
4. John
5. Acts
6. Romans
7. 1 Corinthians
8. 2 Corinthians
9. Galatians
10. Ephesians
11. Philippians
12. Colossians
13. 1 Thessalonians
14. 2 Thessalonians
15. 1 Timothy
16. 2 Timothy
17. Titus
18. Philemon
19. Hebrews
20. James
21. 1 Peter
22. 2 Peter
23. 1 John
24. 2 John
25. 3 John
26. Jude
27. Revelation

What is:

$66 - 39 =$ -----

$39 + 27 =$ -----

~~Genesis 1:1~~

In the beginning God created the heavens and the earth.

~~Genesis 1:27~~

So God created mankind in his own image; in the image of God he created them; male and female he created them.

~~Genesis 9:13~~

I have set my rainbow in the clouds, and it will be the sign of the covenant between me and the earth.

~~John 3:16~~

For God so loved the world that he gave his one and only Son, that whoever believes in him shall not perish but have eternal life.

~~Philippians 4:13~~

I can do all this through him who gives me strength.

~~ 1 Corinthians 16:13-14~~

Be on your guard; stand firm in the faith; be courageous; be strong. Do everything in love.

~~Proverbs 15:5~~

A fool spurns a parent's discipline, but whoever heeds correction shows prudence.

~~Proverbs 15:20~~

A wise son brings joy to his father, but a foolish man despises his mother.

~~Proverbs 1:8~~

Listen, my son, to your father's instruction and do not forsake your mother's teaching.

~~Colossians 3:20~~

Children, obey your parents in everything, for this pleases the Lord.

~~Matthew 19:14~~

Jesus said, "Let the little children come to me, and do not hinder them, for the kingdom of heaven belongs to such as these."

~~Matthew 21:22~~

If you believe, you will receive whatever you ask for in prayer.

~~Mark 10:15~~

Truly I tell you, anyone who will not receive the kingdom of God like a little child will never enter it.

~~Mark 9:37~~

Whoever welcomes one of these little children in my name welcomes me; and whoever welcomes me does not welcome me but the one who sent me.

~~Ephesians 4:32~~

Be kind and compassionate to one another, forgiving each other, just as in Christ God forgave you.

Deuteronomy 31:6
Be strong and courageous.

~Genesis 46:26~~

All those who went to Egypt with Jacob—those who were his direct descendants, not **counting** his sons' wives—numbered sixty-six persons.

One Hundred Chart

Fill in the missing numbers-

1	2	3	4	5		7	8	9	10
11		13	14		16	17		19	20
21	22	23		25	26		28	29	
	32		34	35		37	38		40
41		43	44	45	46	47		49	50
	52	53			56		58		
61	62		64	65		67	68	69	70
71		73	74	75	76	77		79	80
81	82	83			86		88	89	
	92		94	95		97	98		100

spell each number, write the answer on the line

1   *One*

2

3

4

5

6

7

8

9

10

~~**Be kind and helpful to others**~~

Jesus had 12 disciples in the Temple, 3 went to feed the poor, 2 went to heal the sick and 1 went to make your favorite dinner.

How many stayed in the Temple with Jesus? Circle the correct answer:

(a)   7   (b)   6   (c)   4   (d)   9

~~**Genesis 7:2**~~

Take with you 7 pairs- male and female of each animal I have approved for eating and sacrifice and take one pair of each of the others.

How many approved animals did Noah take with him?

(a)   7   (b)   10   (c)   14   (d)   17

~~**Genesis 7:4**~~

Seven days from now I will make the rains pour down on the earth. And it will rain for forty days and forty nights, until I have wiped from the earth all the living things I have created.

If today is Tuesday, when will it rain?

(a) Tuesday   (b) Monday   (c) Saturday   (d) Sunday

~~**Genesis 8:4**~~

Exactly five months from the time the flood began, the boat came to rest on the mountain of Ararat.

If this month is May, when will the boat come to rest on the mountain? Circle the correct answer:

(a) August   (b) October   (c) September (d) November

~~Genesis 8:5~~

Two and a half months later, as the waters continued to go down, other mountain peaks became visible.

Using the information in previous question, what month did the mountain became visible?

(a) October   (b) November   (c) December   (d) January

~~Genesis 8:12~~

He waited another seven days and then released the dove again. This time it did not come back.

If the dove returned on Sunday, when was it released again?

(a) Friday   (b) Sunday   (c) Thursday   (d) Saturday

~~Genesis 8:22~~

As long as the earth remains, there will be planting and harvest, cold and heat, summer and winter, day and night.

Go ahead; test your skills- draw a line to the matching pattern

Cold   Night

Heat   Winter

Day   Summer

Plant some seeds and watch them grow!

~~Genesis 9:18~~

The sons of Noah who came out of the ark were Shem, Ham and Japheth.

Circle the number of sons who came out of the ark with Noah.

(a) 4   (b) 3   (c) 2   (d) 5

~~Genesis 7:12~~

And rain fell on the earth forty days and forty nights.

Go ahead, do the math, add:

$$\begin{array}{r} 40 \\ +\quad 40 \\ \hline \end{array}$$

~~ Psalms 23~~

The Lord is my shepherd, I lack nothing. He makes me lie down in green pastures, he leads me beside quiet waters, he refreshes my soul. He guides me along the right paths for his name's sake. Even though I walk through the darkest valley, I will fear no evil, for you are with me; your rod and your staff, they comfort me.

You prepare a table before me in the presence of my enemies.
You anoint my head with oil; my cup overflows. Surely your goodness and love will follow me all the days of my life, and I will dwell in the house of the Lord forever.

David says the Lord anoints his <u>what</u> with oil?

Head   Hands
Heart   Feet

~~Genesis 7:11, Genesis 9:29~~

Noah was 600 years old when the flood came. He lived for another 350 years.

How old was Noah when he died?

(b) 800   (b) 850   (c) 350   (d) 950

~~Genesis 11:10~~

Two years after the flood, when Shem was 100 years old, he became the father of Arphaxad.

How old was Shem when the flood came?

(c)   100   (b) 102   (c) 98   (d) 110

John 3:6
Humans can reproduce only human life, but the Holy Spirit gives birth to spiritual life.

~~Genesis 11:12~~

When Arphaxad was 35 years old, he became the father of Shelah. He lived another 403 years.

How old was Arphaxad when he died?

$$35$$
$$+ \underline{403}$$

~~Genesis 33:1~~

Then Jacob looked up and saw Esau coming with his 400 men.

If Esau sends half of his men back home, how many men stayed with Esau?

(a) 100   (b) 150   (c) 200   (d) 250

~~Genesis 35:28~~

Isaac lived for 180 years.

How old was he 20 years earlier?

180 minus 20 = _____

~~Genesis 4:1~~

Adam and Eve had two sons, Cain and Abel.

If they had three more sons, how many sons will they have in total?

(a) 2   (b) 5   (c) 4   (d) 6

~~Matthew 6:33~~

But seek first his kingdom and his righteousness; and all these things will be given to you as well.

~~Genesis 46:26~~

The total number of Jacob's direct descendants who went with him to Egypt was 66.

If 23 were females, how many were males?

$$\begin{array}{r} 66 \\ -\ 23 \\ \hline \end{array}$$

~~Exodus 20:9~~

You have six days each week for your ordinary work, but the seventh day is the Sabbath day.

How many days are in a week?

(a) 6   (b) 7   (c) 5   (d) 8

~~Numbers 3:46~~

There are 273 more firstborn sons of Israel than there are Levites.

If there are 210 Levites, how many more firstborn sons of Israel are there?

(a) ____ = 273 + 210   (c) 483 − 273 = ____

(b) ____ = 273 − 210   (d) 483 − 210 = ____

~~Matthew 10:2~~

These are the names of Jesus' apostles-

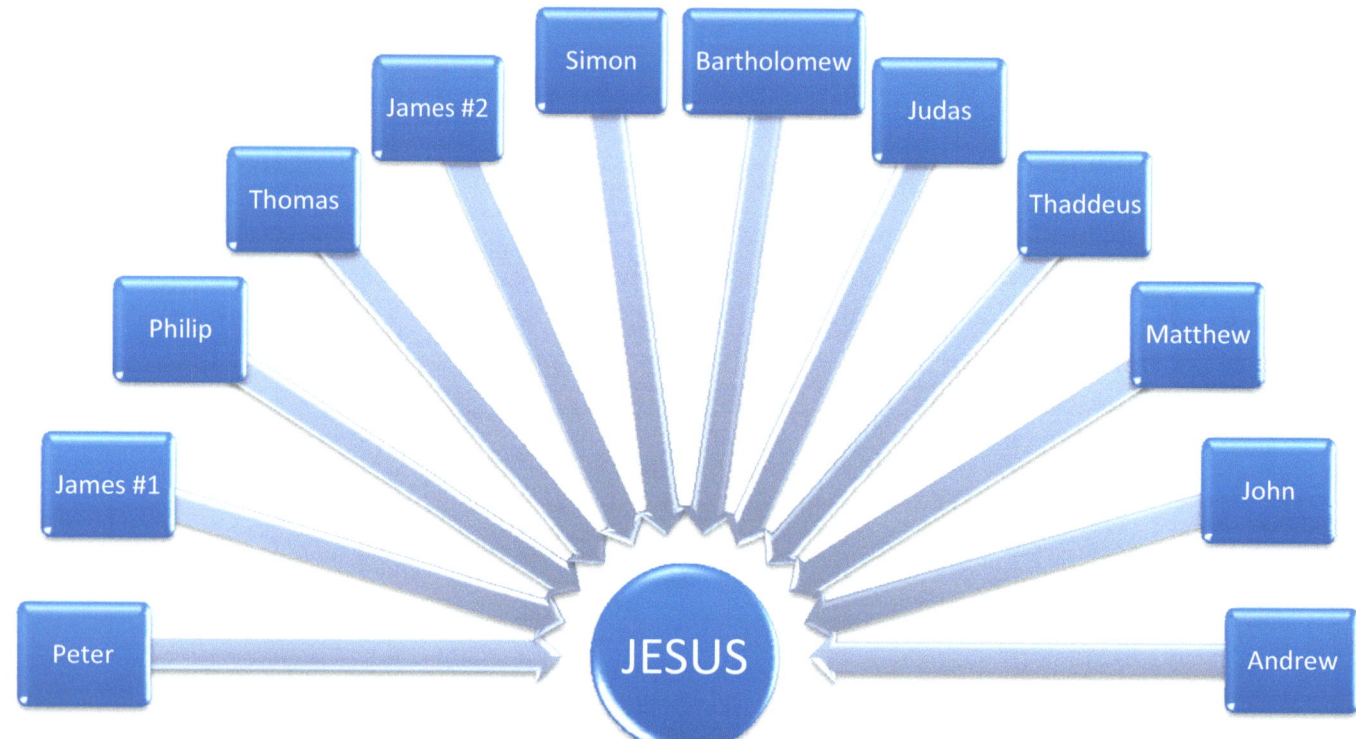

(a) How many apostles did Jesus have altogether _____

(b) How many of their names begin with the letter **J**? ____

(c) How many of their names begin with the letter **T**? ____

~~Matthew 17:1~~

Six days later Jesus took Peter and the two brothers, James and John and led the up to the high mountains to be alone.

(a)   How many people went up the mountains with Jesus? ____

(b)   When did they go up to the high mountains? ____

Multiplications –Matthew, Mark and Luke fish in numbers!

~~Mark 6:41~~

And when he had taken the **five** loaves and the **two fishes**, he looked up to heaven, and blessed, and breaks the loaves, and gave *them* to his disciples to set before them; and the **two fishes** divided he among them all.

~~Matthew 14:17~~

And they say unto him, we have here but **five** loaves, and **two fishes**.

5 x 2 = ____   5 + 2 = ____   5 - 2 = ____

~~Luke 5:9~~

For he was astonished, and all that were with him, at the draught of the **fishes** which they had taken:

~~Matthew 15:34~~

And Jesus saith unto them, how many loaves have ye? And they said, **seven,** and a few little **fishes**.

If few equal 5, what is?

7 x 5 = ____   7 + 5 = ____   7 – 5 = ____

~~Mark 6:38~~

He saith unto them, how many loaves have ye? Go and see. And when they knew, they say, **five**, and **two** **fishes**.

If 5 x 2 = 10, write in the missing number:

5 + ___ = 10

~~Exodus 25:23~~

Then make a table of wood 36 inches long and 18 inches wide.

Calculate the following:

36 x 18 = ____   36 + 18 = ____

36 − 18 = ____   18 + 18 = ____

~~Free Space~~

Counting by 2s

Color every other number *"red"*, begin with number 2

1	2	3	4	5	6	7	8	9	10
11	12	13	14	15	16	17	18	19	20
21	22	23	24	25	26	27	28	29	30
31	32	33	34	35	36	37	38	39	40
41	42	43	44	45	46	47	48	49	50
51	52	53	54	55	56	57	58	59	60
61	62	63	64	65	66	67	68	69	70
71	72	73	74	75	76	77	78	79	80
81	82	83	84	85	86	87	88	89	90
91	92	93	94	95	96	97	98	99	100

Write down all the even numbers you have colored *"red"*.

2 ___ ___ ___ ___ ___ ___ ___ ___ ___

___ ___ ___ ___ ___ ___ ___ ___ ___ ___

___ ___ ___ ___ ___ ___ ___ ___ ___ ___

___ ___ ___ ___ ___ ___ ___ ___ ___ ___

___ ___ ___ ___ ___ ___ ___ ___ ___ 100

~~**Genesis 7:9**~~

There went in **two** and **two** unto Noah into the ark, the male and the female, as God had commanded Noah.

Draw your favorite pair of animals here-

(a)   2 + 2 + 2 = _____

(b)   2 x 2 = _____

(c)   2 x 3 = _____

(d)   2 + 2 + 2 + 2 = _____

~~**Revelation 11:4**~~

They are the **two** olive trees and the **two** lampstands, and they stand before the Lord of the earth.

Fill in the missing numbers:

(e)   2, 4, 6 __, 10, __, 14, __, __

 2    __    6   __   __

(f)   20, __, 24 __, __, 30, __, __, __

Color every fifth number **"blue"**, begin with number 5

1	2	3	4	5	6	7	8	9	10
11	12	13	14	15	16	17	18	19	20
21	22	23	24	25	26	27	28	29	30
31	32	33	34	35	36	37	38	39	40
41	42	43	44	45	46	47	48	49	50
51	52	53	54	55	56	57	58	59	60
61	62	63	64	65	66	67	68	69	70
71	72	73	74	75	76	77	78	79	80
81	82	83	84	85	86	87	88	89	90
91	92	93	94	95	96	97	98	99	100

Write down all the numbers you have colored **"blue"**

5 ___ ___ ___ 25

___ ___ ___ ___ ___

___ ___ ___ ___ ___

___ ___ ___ ___ 100

Fun with 5s

~~Leviticus 27:5~~

And if *it be* from **five** years old even unto **twenty** years old, then thy estimation shall be of the male **twenty** shekels, and for the female **ten** shekels.

Complete the following:

(a) 5 + 5 = _____

(d) 5 + 5 + 5 = _____

(b) 5 + 5 + 5 + 5 = _____
_____

(e) 5 + 5 + 5 + 5 + 5 =

(c) 5 x 4 = _____

(f) 5 + 0 = _____

Fill in the missing numbers:

(g) 25, ___, ___, 40, 45, 50, ___, ___

(h) 100, ____, ____, 115, 120, ____, 130, ____, ____, ____

(i) 65, ____, 75, ____, ____, ____, ____, ____, ____

~~**Genesis 47:14**~~

Joseph collected all the **money** that was to be found in Egypt and Canaan in payment for the grain they were buying, and he brought it to Pharaoh's palace.

Try these:

(1)   5 tens and 9 ones = _____

(2)   Five hundred ___tens ___ones = 519

$$\begin{array}{r} 500 \\ 95 \\ +5 \\ \hline \end{array}$$

(3)   6 tens and 7 ones = ___

$$\begin{array}{r} 700 \\ 70 \\ +7 \\ \hline \end{array}$$

(4)   7 hundreds ___tens ___ones = 777

(5)   7 + 7 + 7= ___

(6)   8 tens and 9 ones= ___

(7)   66 = ___ tens and ___ones

(8)   27 = ___ tens and ___ ones

(9)   84 = ___ tens and ___ones

(10)   94 = 9 tens and ___ones

(11)   68 = ___tens and ___ones

(12)   99 = ___ tens and 9 ones

Show your work here:

(1)   Take 11 from 57 = _____

(2)   Add 13, 10 and 9 = _____

(3)   Seventy-seven minus thirty two = _____

(4)   Increase 34 by 12 = _____

(5)   Reduce 110 by 17= _____

(6)   By how much is 66 greater than 50 = _____

(7)   By how much is 10 less than 51 = _____

(8)   What is the total of 16, 23 and 11 = _____

(9)   Subtract 16 from 59 = _____

(1)   77 =_____+_____

(2)   603 =_____+___+____

(3)   900 =_____+_____+____

(4)   ____+_____+_____+_____= 800

(5)   950 =_____+_____+_____

Write 110 in different ways, example 90 + 20 =110

a)   ___ + ___ + ___ =110

b)   ___ + ___ =110

c)   ___ - ___ =110

d)   ___ + ___ + ___ + ___ =110

e)   10 x ___ =110

f)   220 – 110 = ___

g)   55 x 2 = ___

Write the value of each coin on the line below each coin and the total in the box.

_____ ¢   _____ ¢   _____ ¢   _____ ¢   _____ ¢   _____ ¢

\[____ ¢ \]

_____ ¢   _____ ¢   _____ ¢

\[____ ¢ \]

_____ ¢   _____ ¢   _____ ¢   _____ ¢   _____ ¢

\[____ ¢ \]

_____ ¢   _____ ¢   _____ ¢   _____ ¢

\[____ ¢ \]

_____ ¢   _____ ¢   _____ ¢

\[____ ¢ \]

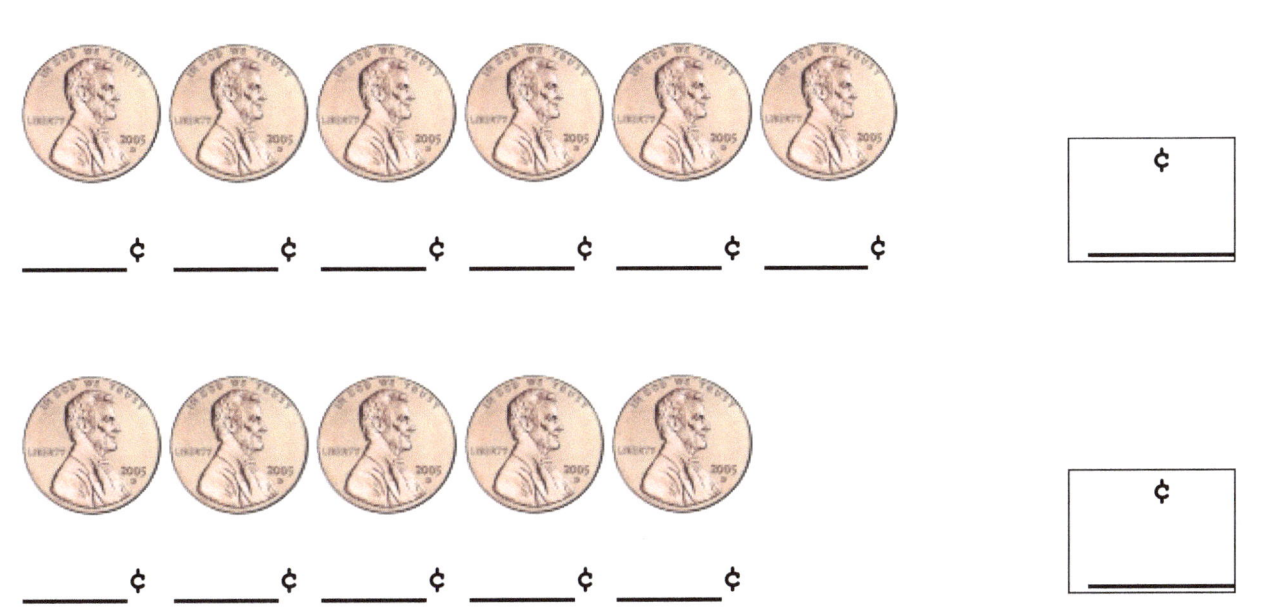

_____ ¢ _____ ¢ _____ ¢ _____ ¢ _____ ¢ _____ ¢ | ¢
                                                      |_____

_____ ¢ _____ ¢ _____ ¢ _____ ¢ _____ ¢ | ¢
                                                      |_____

Finding Values

Count the coins. Write the value of each coin on the line and the total in the box.

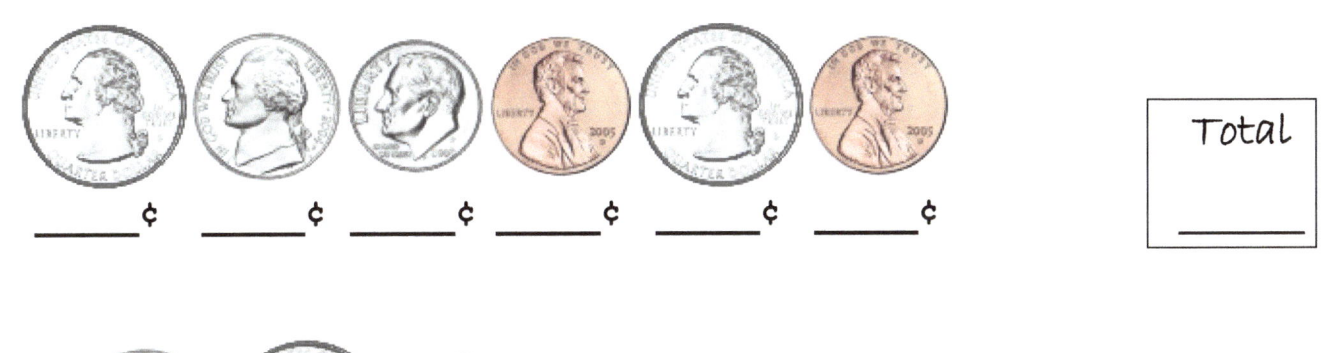

_____ ¢ _____ ¢ _____ ¢ _____ ¢ _____ ¢ _____ ¢ | Total
                                                      |_____

_____ ¢ _____ ¢ _____ ¢ | Total
                                                      |_____

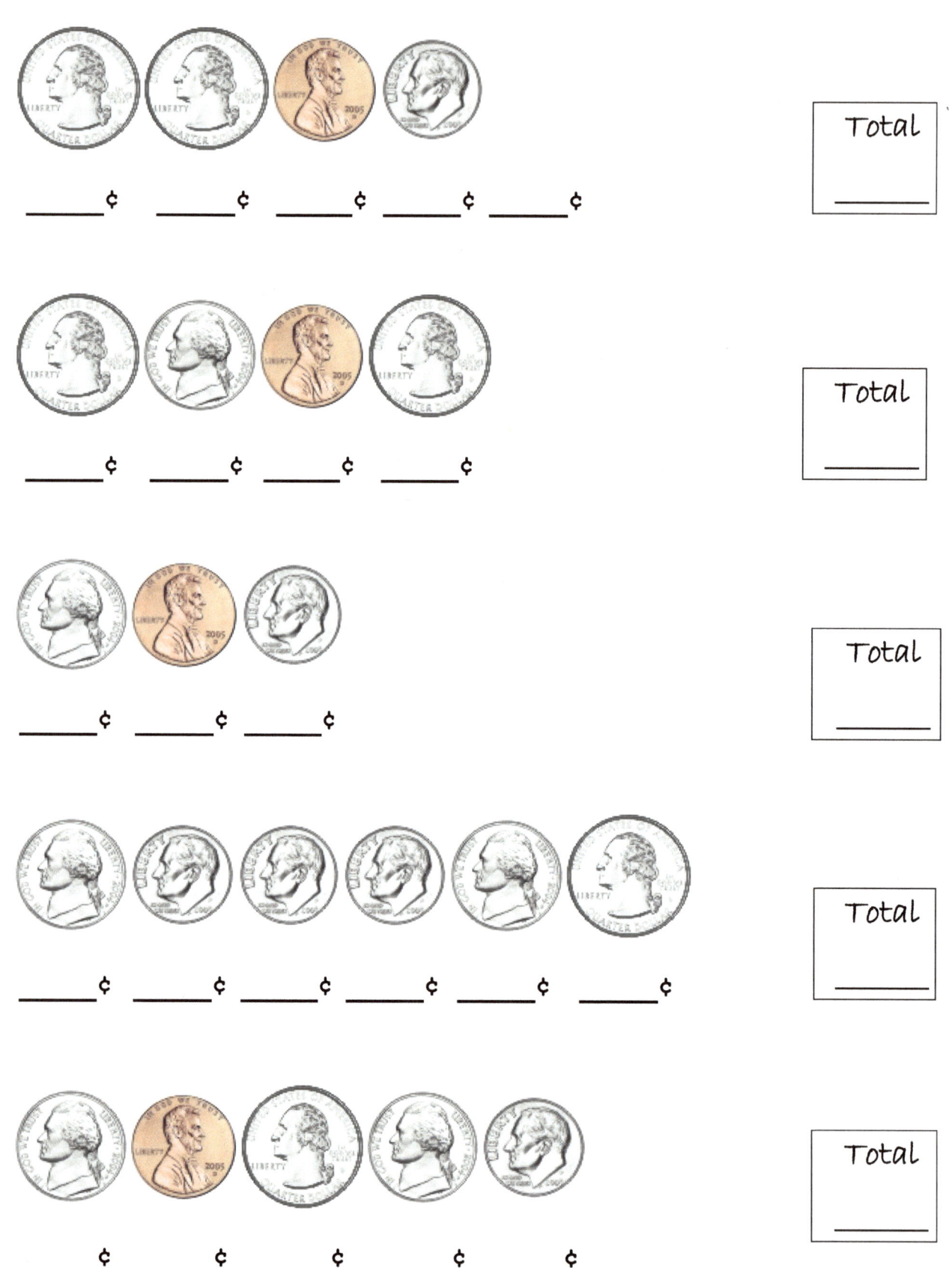

_____¢ _____¢ _____¢ _____¢ _____¢

Total

_____

_____¢ _____¢ _____¢ _____¢

Total

_____

_____¢ _____¢ _____¢

Total

_____

_____¢ _____¢ _____¢ _____¢ _____¢ _____¢

Total

_____

_____¢ _____¢ _____¢ _____¢ _____¢

Total

_____

Count the coins and write the total amount in the box

Bonus Question:

You went to the store to buy snacks; you picked up 1 bag of popcorn for 50 cents, 2 lollipops for 40 cents each. You gave the cashier $5.00. How much change will you receive from the cashier?

Count the coins and write the total amount in the octagon.

Write the number of coins for the total amount.

 66¢   ____ quarters, ___ dimes, ___ nickels, ____ pennies

 55¢   ____quarters, ____ dimes, ____ nickels

 96¢   ____ quarters, ____ dimes, ____ pennies

 76¢   ____ quarters, ____ nickels, ____ pennies

 26¢   ___ dimes, ____ nickels, ____ pennies

 95¢   ____ quarters, ____ dimes

 27¢   ____ nickels, ____ pennies

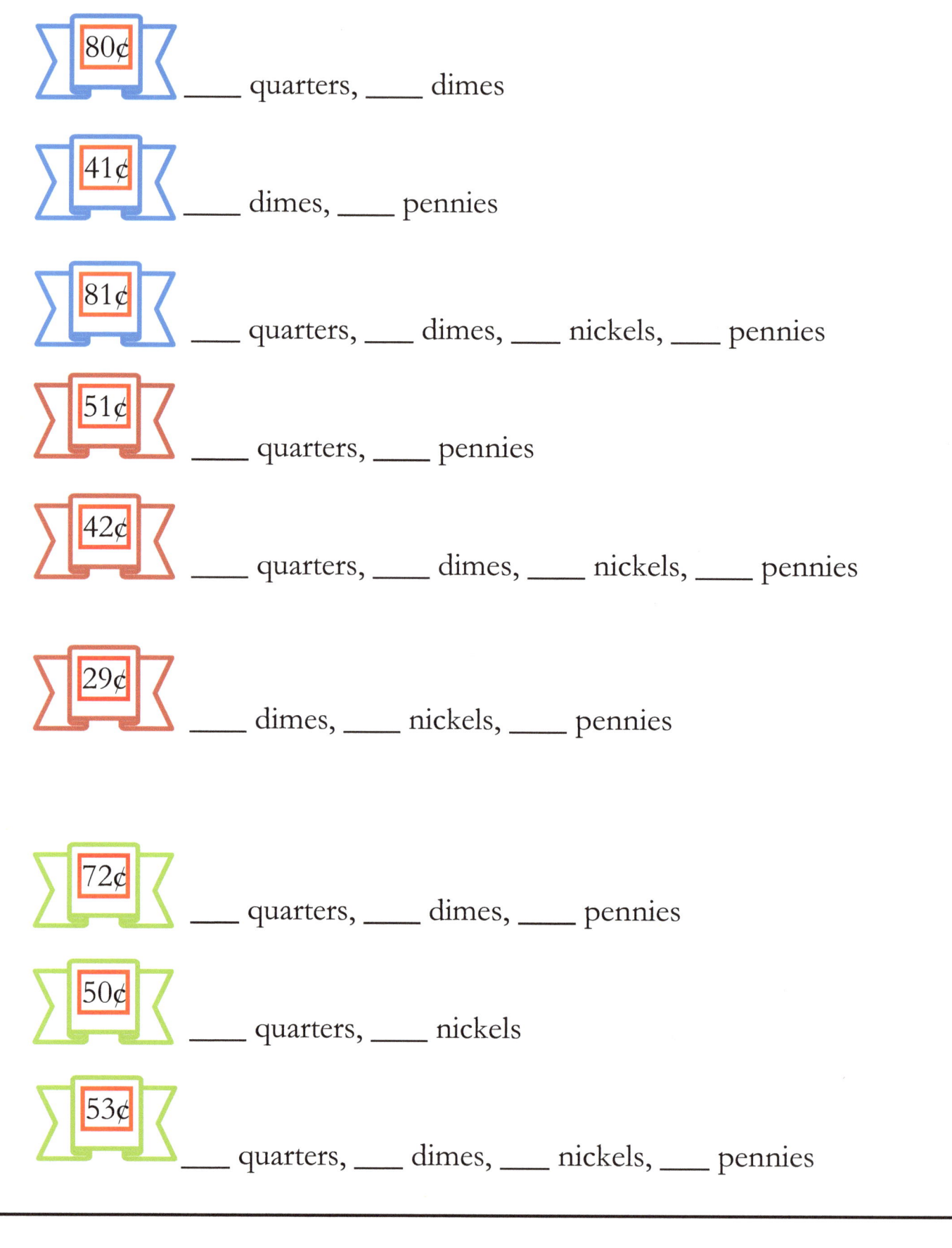

80¢ ____ quarters, ____ dimes

41¢ ____ dimes, ____ pennies

81¢ ____ quarters, ____ dimes, ____ nickels, ____ pennies

51¢ ____ quarters, ____ pennies

42¢ ____ quarters, ____ dimes, ____ nickels, ____ pennies

29¢ ____ dimes, ____ nickels, ____ pennies

72¢ ____ quarters, ____ dimes, ____ pennies

50¢ ____ quarters, ____ nickels

53¢ ____ quarters, ____ dimes, ____ nickels, ____ pennies

You did a great job; now you can count all the money you saved up for your favorite toy!

Write the values-learn how to count the money in your piggy bank.

Forty dollars and ten cents

$40.10

Seventy five dollars and thirty cents

$_____

Twenty dollars and fifty cents

$_____

Eighty two dollars and fifty three cents

$_____

Five hundred dollars and fourteen cents

$_____

Two hundred and fifty nine dollars

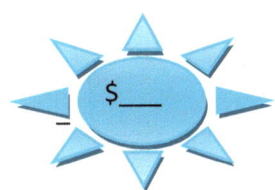
$_____

Sixteen dollars and two cents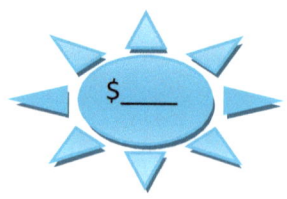

Eight dollars and forty eight cents

Eleven dollars and twenty two cents

Thirty one dollars and three cents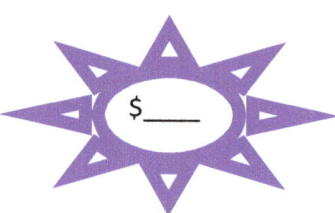

A time to be born and a time to die,
A time to plant and a time to uproot,
A time to kill and a time to heal,
A time to tear down and a time to build,
A time to weep and a time to laugh,
A time to mourn and a time to dance,
A time to scatter stones and a time to gather them,
A time to embrace and a time to refrain from embracing,
A time to search and a time to give up,
A time to keep and a time to throw away,
A time to tear and a time to mend,
A time to be silent and a time to speak,
A time to love and a time to hate,
A time for war and a time for peace.

Are we there yet? What time is it? It's not bed time yet!

Can you tell the time? Go ahead and test your skills

1:50

___ : ___

10:00

___ : ___

~~Genesis 4:3~~

In the course of **time** Cain brought some of the fruits of the soil as an offering to the Lord.

4:30

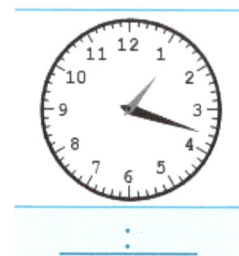

_:_

12:00

_:_

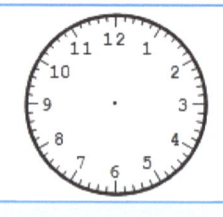

12:45

_:_

6:30

_:_

1:30

_:_

7:45

YOU DID A FANTASTIC JOB! GREAT WORK!

2:45

___ : ___

3:50

___ : ___

3:45

___ : ___

4:30

___ : ___

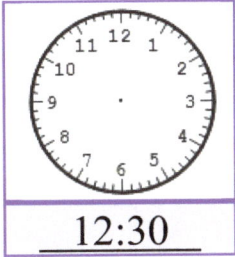

12:30

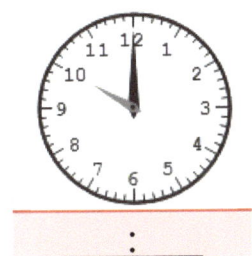

___ : ___

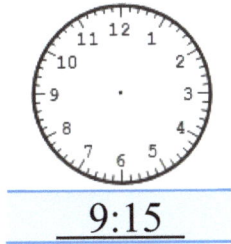

9:15

___ : ___

9:00

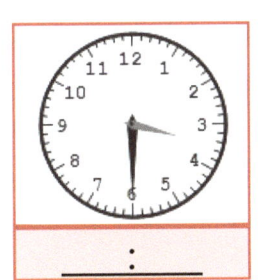

___ : ___

1:30

___ : ___

Find the missing numbers

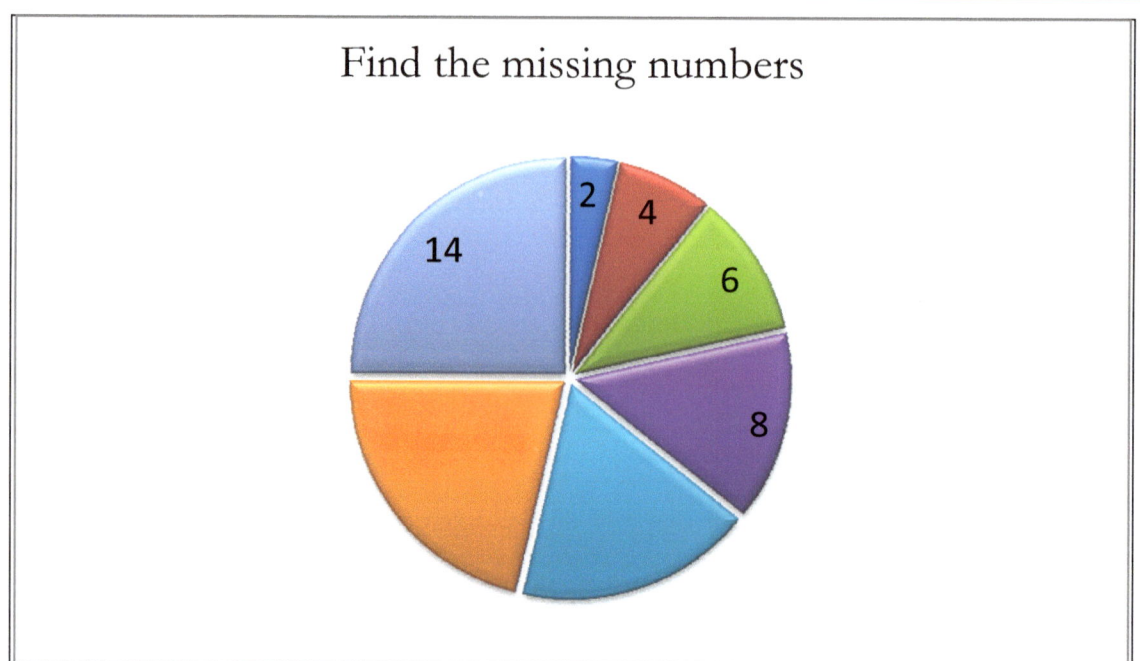

Counting by 3s, find the missing numbers

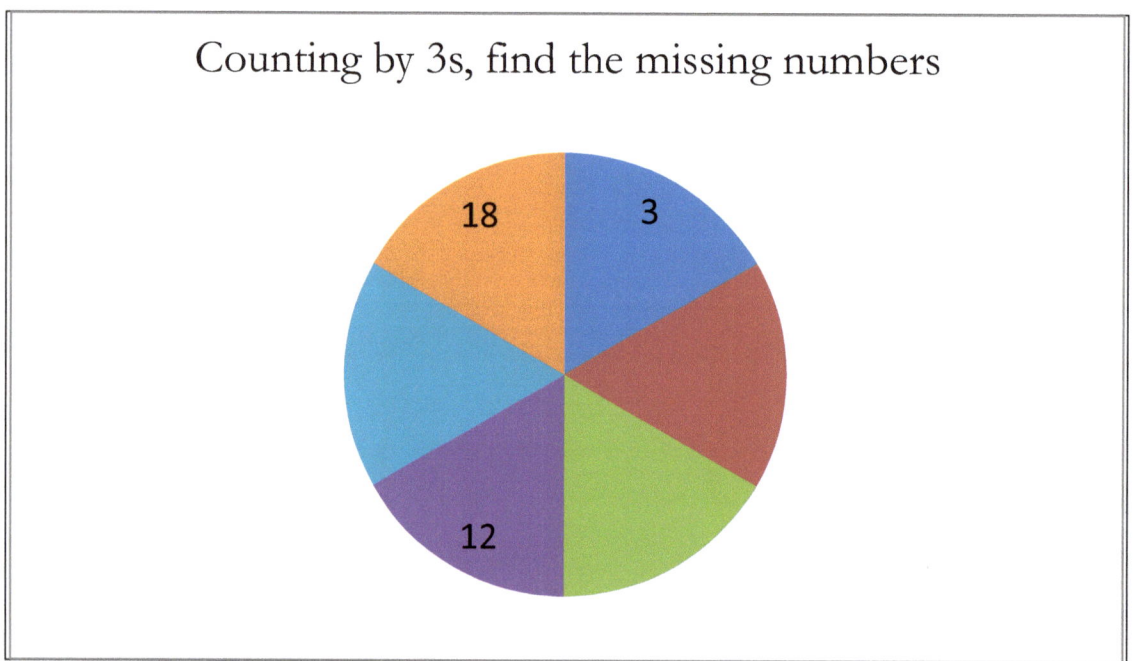

3 x 3 = _____   3 x 1 = _____   3 x 4 = _____

3 x 2 = _____   3 x 5 = _____   3 x 6 = _____

Can you find the missing numbers?

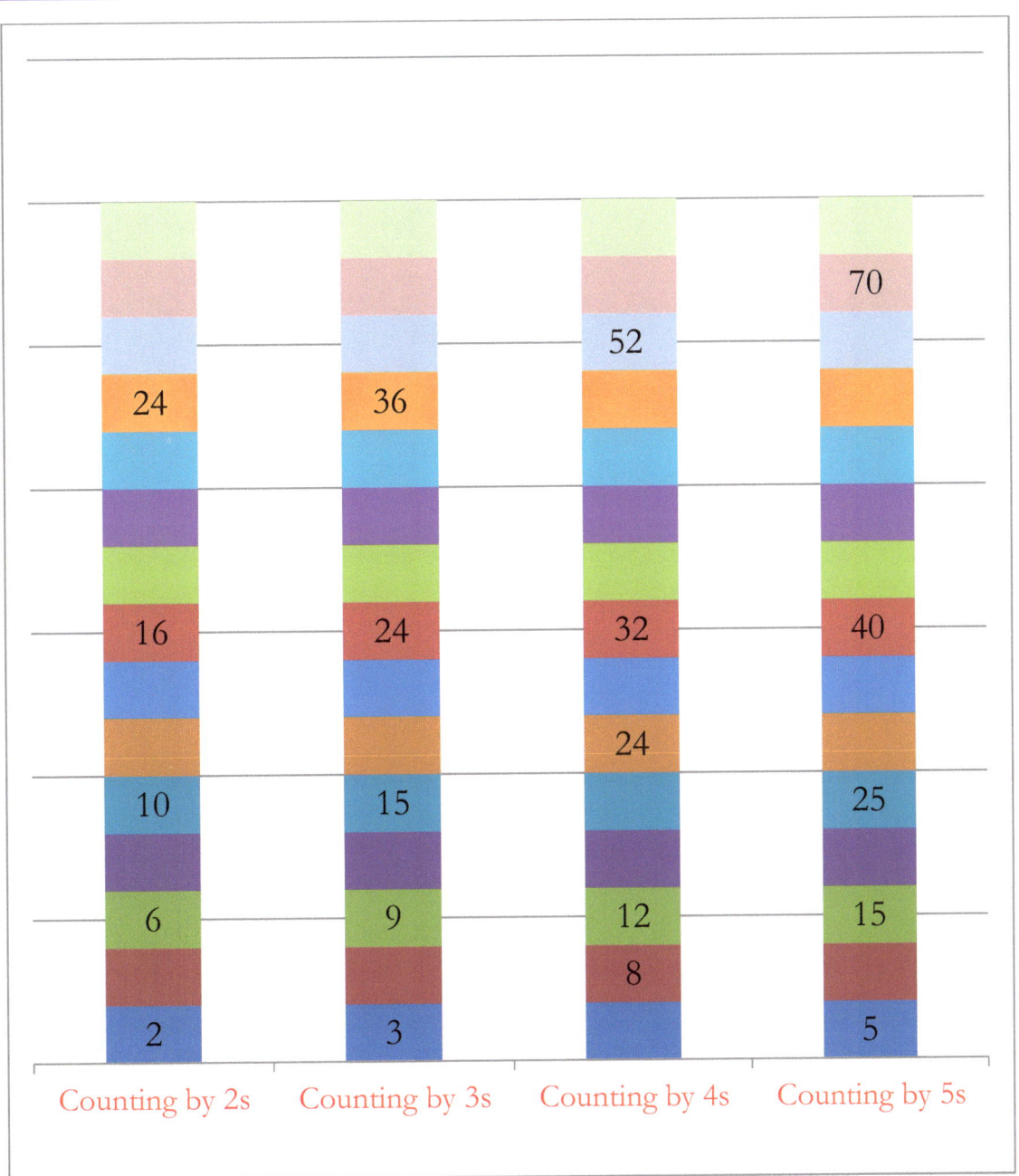

Counting by 2s	Counting by 3s	Counting by 4s	Counting by 5s
24	36	52	70
16	24	32	40
		24	
10	15		25
6	9	12	15
		8	
2	3		5

281 + 162	158 + 224	245 + 139
200 + 294	505 + 214	156 + 142
109 + 106	158 + 163	199 + 148
324 + 148	231 + 128	520 + 260
168 + 192	212 + 158	216 + 106

$$
\begin{array}{r} 414 \\ + 606 \\ \hline \end{array}
\qquad
\begin{array}{r} 850 \\ + 520 \\ \hline \end{array}
\qquad
\begin{array}{r} 324 \\ + 814 \\ \hline \end{array}
$$

$$
\begin{array}{r} 136 \\ + 127 \\ \hline \end{array}
\qquad
\begin{array}{r} 640 \\ + 109 \\ \hline \end{array}
\qquad
\begin{array}{r} 207 \\ + 306 \\ \hline \end{array}
$$

$$
\begin{array}{r} 201 \\ + 114 \\ \hline \end{array}
\qquad
\begin{array}{r} 168 \\ + 276 \\ \hline \end{array}
\qquad
\begin{array}{r} 430 \\ + 166 \\ \hline \end{array}
$$

$$
\begin{array}{r} 185 \\ + 225 \\ \hline \end{array}
\qquad
\begin{array}{r} 190 \\ + 146 \\ \hline \end{array}
\qquad
\begin{array}{r} 534 \\ + 218 \\ \hline \end{array}
$$

$$
\begin{array}{r} 524 \\ + 346 \\ \hline \end{array}
\qquad
\begin{array}{r} 742 \\ + 341 \\ \hline \end{array}
\qquad
\begin{array}{r} 456 \\ + 860 \\ \hline \end{array}
$$

```
   354            249            128
+  132         +  748         +  855


   292            567            508
+  778         +  295         +  384


   485            731            947
+  353         +  178         +  205


   527            921            345
+  576         +  374         +  162


   837            489            508
+  909         +  860         +  866
```

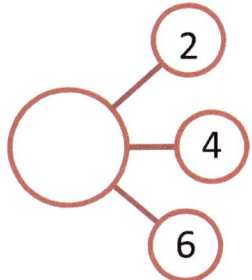

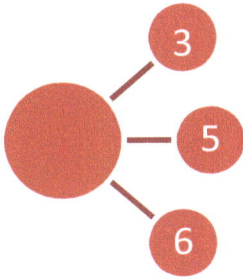

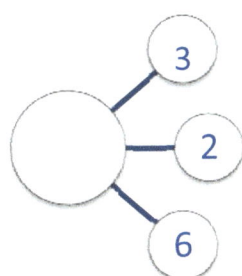

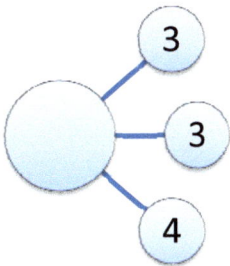

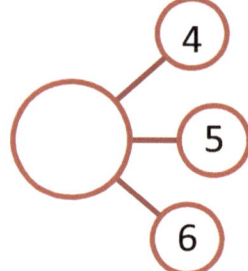

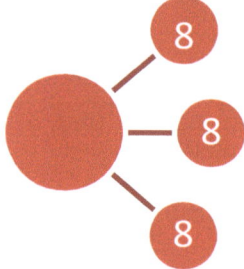

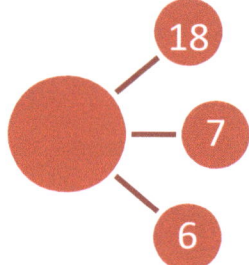

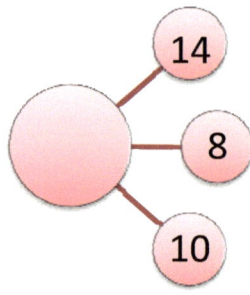

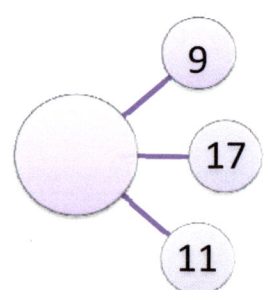

$$205 + 15$$

$$220 + 14$$

$$301 + 22$$

$$410 + 42$$

$$530 + 2$$

$$48 + 4$$

$$23 + 3$$

$$24 + 2$$

$$22 + 4$$

$$21 + 4$$

$$35 + 3$$

$$41 + 2$$

$$53 + 2$$

$$472 + 2$$

$$32 + 3$$

$$42 + 11$$

$$55 + 12$$

$$71 + 13$$

	80		60		83
+	12	+	5	+	3

	73		84		80
+	42	+	26	+	14

	70		82		90
+	33	+	48	+	43

	90		70		93
+	22	+	5	+	3

	43		64		90
+	42	+	26	+	14

	70		62		50
+	23	+	47	+	43

94 + 2	72 + 6	115 + 4
115 + 14	114 + 3	109 + 32
111 + 5	115 + 25	112 + 2
105 + 2	106 + 4	118 + 22
124 + 12	106 + 24	124 + 12
105 + 24	109 + 32	119 + 34

$$107 + 5$$

$$106 + 15$$

$$118 + 22$$

$$135 + 3$$

$$107 + 13$$

$$108 + 22$$

$$114 + 3$$

$$114 + 33$$

$$129 + 44$$

$$116 + 53$$

$$238 + 169$$

$$184 + 176$$

$$214 + 23$$

$$104 + 30$$

$$329 + 54$$

$$216 + 43$$

$$338 + 149$$

$$384 + 166$$

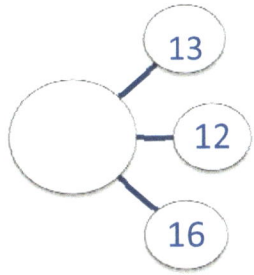

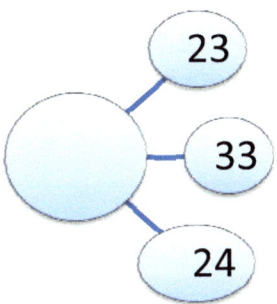

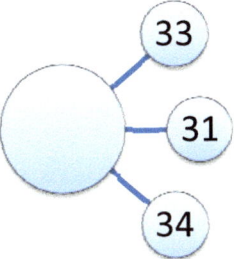

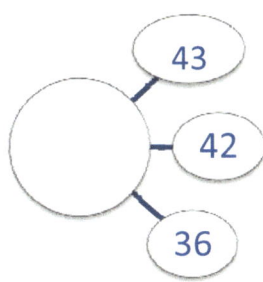

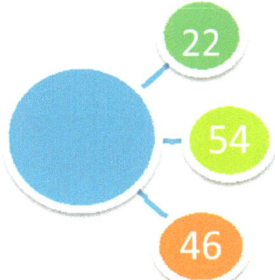

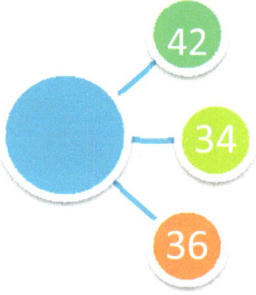

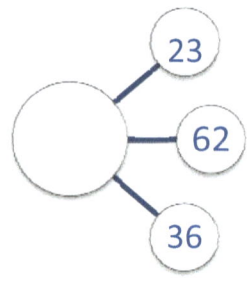

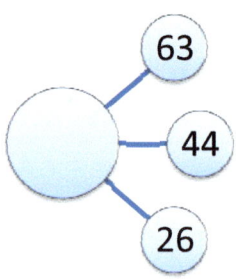

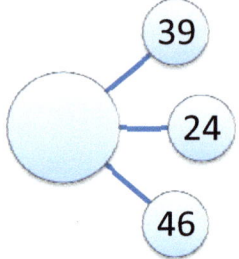

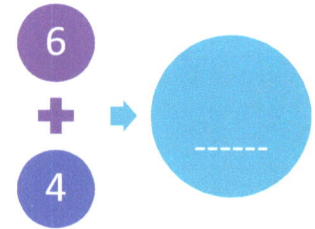

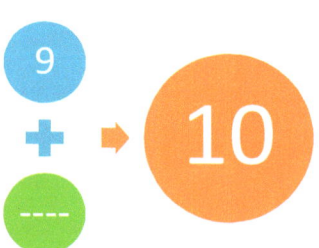

5 + 5 = ___

7 + 3 = ___

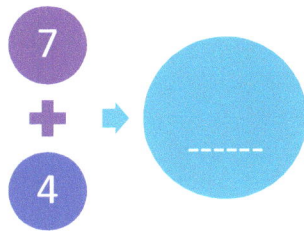

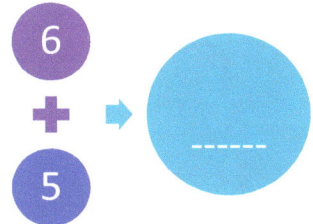

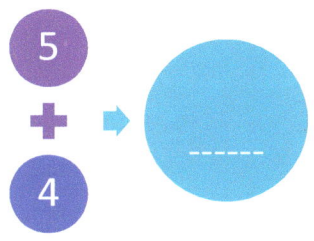

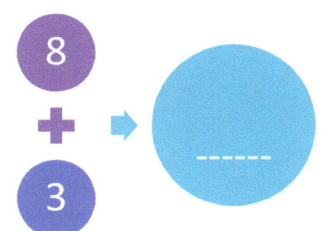

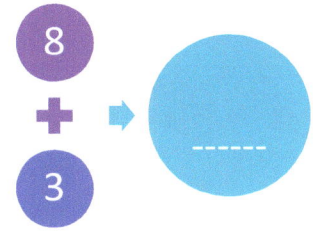

$3 + 3 =$ ____

$4 + 4 =$ ____

$6 + 4 =$ ____

$5 + 7 =$ ____

Subtraction- Solve for the missing numbers-

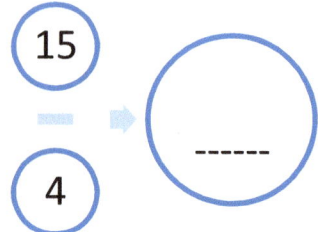

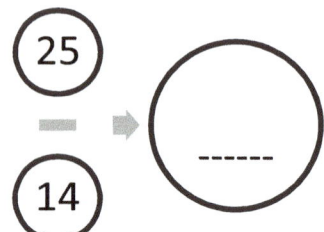

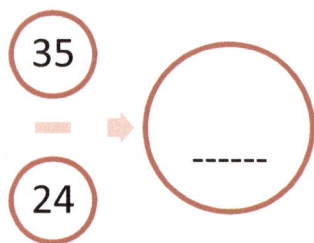

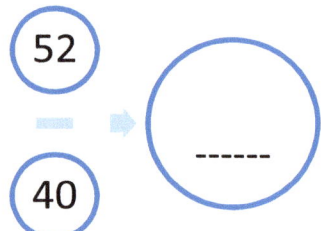

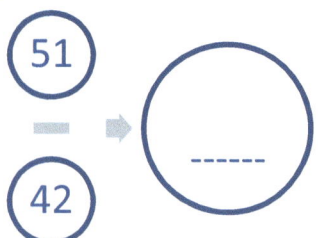

55 − 34 = _____

75 − 48 = _____

59 − 14 = _____

57 − 40 = _____

85 − 64 = _____

95 − 47 = _____

680	248	220
− 145	− 202	− 144

219	335	415
− 147	− 254	− 201

559	817	737
− 201	− 412	− 449

513	470	372
− 125	− 217	− 135

1042	1155	2071
− 810	− 412	− 375

910 − 315	705 − 350	882 − 311
930 − 437	958 − 419	932 − 279
813 − 230	918 − 212	822 − 249
721 − 510	422 − 259	302 − 232
516 − 273	409 − 312	433 − 219
324 − 304	615 − 249	721 − 416

807 − 258	706 − 415	818 − 422
315 − 53	430 − 42	505 − 249
535 − 203	707 − 313	808 − 452
419 − 101	318 − 202	419 − 230
205 − 15	220 − 14	301 − 22

```
    410           53            61
  -  42         -   4         -  19
```

```
     23           24            22
  -    3         -   2         -   4
```

```
    500          612           801
  - 180         -  26         - 314
```

```
    710          882           290
  - 330         - 480         - 243
```

```
    530           80            73
  - 201         -  12         -  14
```

```
   53          472          32
-  12       -  217       -  13


   42           55          71
-  10       -   12       -  33


  610           70         810
-  15       -   33       -  11


  950          938         132
-  43       -   41       -  27


  813          914         722
-  23       -   22       -  24
```

~~Genesis 47:21~~

And Joseph **reduced** the people to servitude, from one end of Egypt to the other.

$$\begin{array}{r} 121 \\ -\ \ 5 \\ \hline \end{array} \qquad \begin{array}{r} 125 \\ -\ 25 \\ \hline \end{array} \qquad \begin{array}{r} 102 \\ -\ 32 \\ \hline \end{array}$$

$$\begin{array}{r} 116 \\ -\ 73 \\ \hline \end{array} \qquad \begin{array}{r} 109 \\ -\ 32 \\ \hline \end{array} \qquad \begin{array}{r} 133 \\ -\ 12 \\ \hline \end{array}$$

$$\begin{array}{r} 124 \\ -\ \ 3 \\ \hline \end{array} \qquad \begin{array}{r} 115 \\ -\ \ 2 \\ \hline \end{array} \qquad \begin{array}{r} 116 \\ -\ \ 4 \\ \hline \end{array}$$

$$\begin{array}{r} 107 \\ -\ 25 \\ \hline \end{array} \qquad \begin{array}{r} 106 \\ -\ 15 \\ \hline \end{array} \qquad \begin{array}{r} 118 \\ -\ 22 \\ \hline \end{array}$$

$$\begin{array}{r} 113 \\ -53 \\ \hline \end{array} \qquad \begin{array}{r} 124 \\ -12 \\ \hline \end{array} \qquad \begin{array}{r} 105 \\ -24 \\ \hline \end{array}$$

$$\begin{array}{r} 135 \\ -33 \\ \hline \end{array} \qquad \begin{array}{r} 107 \\ -13 \\ \hline \end{array} \qquad \begin{array}{r} 108 \\ -22 \\ \hline \end{array}$$

$$\begin{array}{r} 119 \\ -44 \\ \hline \end{array} \qquad \begin{array}{r} 119 \\ -22 \\ \hline \end{array} \qquad \begin{array}{r} 219 \\ -23 \\ \hline \end{array}$$

Use this space to show your work:

Subtraction- Solve for the missing number.

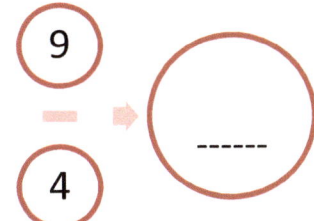

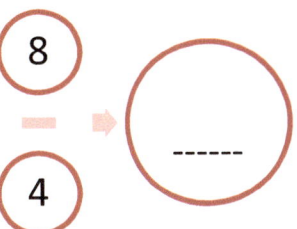

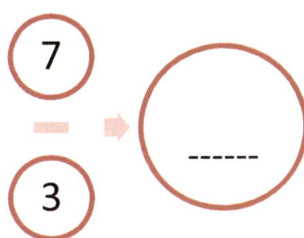

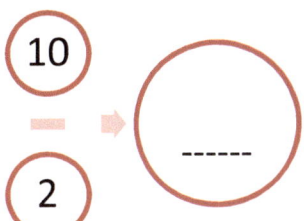

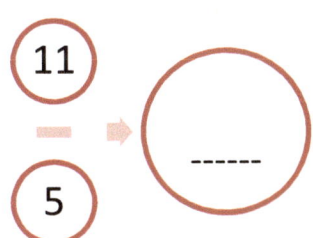

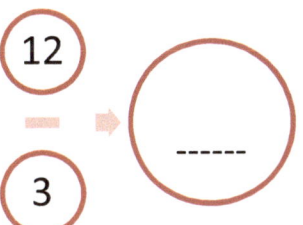

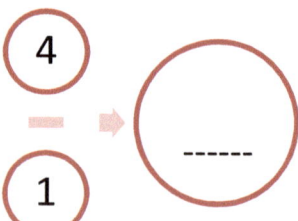

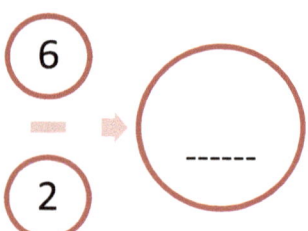

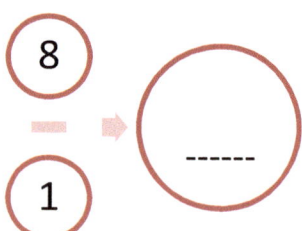

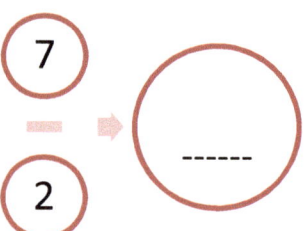

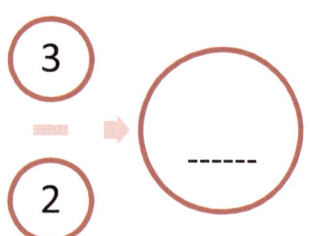

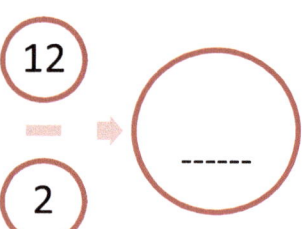

~~Ezekiel 44:23~~

They are to teach my people the **difference** between the holy and the common and show them how to distinguish between the unclean and the clean.

630 - 357	871 - 186	510 - 129
463 - 357	727 - 618	512 - 386
646 - 291	741 - 270	435 - 154
577 - 287	613 - 439	420 - 171
945 - 286	336 - 256	469 - 263
827 - 536	828 - 207	614 - 492

~~**Exodus 6:26**~~

It was this Aaron and Moses to whom the Lord said, "Bring the Israelites out of Egypt by their **division**s."

Division Facts

By 1

$0 \div 1 = 0$
$1 \div 1 = 1$
$2 \div 1 = 2$
$3 \div 1 = 3$
$4 \div 1 = 4$
$5 \div 1 = 5$
$6 \div 1 = 6$
$7 \div 1 = 7$
$8 \div 1 = 8$
$9 \div 1 = 9$
$10 \div 1 = 10$
$11 \div 1 = 11$
$12 \div 1 = 12$

By 2

$0 \div 2 = 0$
$2 \div 2 = 1$
$4 \div 2 = 2$
$6 \div 2 = 3$
$8 \div 2 = 4$
$10 \div 2 = 5$
$12 \div 2 = 6$
$14 \div 2 = 7$
$16 \div 2 = 8$
$18 \div 2 = 9$
$20 \div 2 = 10$
$22 \div 2 = 11$
$24 \div 2 = 12$

By 3

$0 \div 3 = 0$
$3 \div 3 = 1$
$6 \div 3 = 2$
$9 \div 3 = 3$
$12 \div 3 = 4$
$15 \div 3 = 5$
$18 \div 3 = 6$
$21 \div 3 = 7$
$24 \div 3 = 8$
$27 \div 3 = 9$
$30 \div 3 = 10$
$33 \div 3 = 11$
$36 \div 3 = 12$

By 4

$0 \div 4 = 0$
$4 \div 4 = 1$
$8 \div 4 = 2$
$12 \div 4 = 3$
$16 \div 4 = 4$
$20 \div 4 = 5$
$24 \div 4 = 6$
$28 \div 4 = 7$
$32 \div 4 = 8$
$36 \div 4 = 9$
$40 \div 4 = 10$
$44 \div 4 = 11$
$48 \div 4 = 12$

By 5

$0 \div 5 = 0$
$5 \div 5 = 1$
$10 \div 5 = 2$
$15 \div 5 = 3$
$20 \div 5 = 4$
$25 \div 5 = 5$
$30 \div 5 = 6$
$35 \div 5 = 7$
$40 \div 5 = 8$
$45 \div 5 = 9$
$50 \div 5 = 10$
$55 \div 5 = 11$
$60 \div 5 = 12$

By 6

$0 \div 6 = 0$
$6 \div 6 = 1$
$12 \div 6 = 2$
$18 \div 6 = 3$
$24 \div 6 = 4$
$30 \div 6 = 5$
$36 \div 6 = 6$
$42 \div 6 = 7$
$48 \div 6 = 8$
$54 \div 6 = 9$
$60 \div 6 = 10$
$66 \div 6 = 11$
$72 \div 6 = 12$

Division Facts

By 7

0 ÷ 7 = 0
7 ÷ 7 = 1
14 ÷ 7 = 2
21 ÷ 7 = 3
28 ÷ 7 = 4
35 ÷ 7 = 5
42 ÷ 7 = 6
49 ÷ 7 = 7
56 ÷ 7 = 8
63 ÷ 7 = 9
70 ÷ 7 = 10
77 ÷ 7 = 11
84 ÷ 7 = 12

By 8

0 ÷ 8 = 0
8 ÷ 8 = 1
16 ÷ 8 = 2
24 ÷ 8 = 3
32 ÷ 8 = 4
40 ÷ 8 = 5
48 ÷ 8 = 6
56 ÷ 8 = 7
64 ÷ 8 = 8
72 ÷ 8 = 9
80 ÷ 8 = 10
88 ÷ 8 = 11
96 ÷ 8 = 12

By 9

0 ÷ 9 = 0
9 ÷ 9 = 1
18 ÷ 9 = 2
27 ÷ 9 = 3
36 ÷ 9 = 4
45 ÷ 9 = 5
54 ÷ 9 = 6
63 ÷ 9 = 7
72 ÷ 9 = 8
81 ÷ 9 = 9
90 ÷ 9 = 10
99 ÷ 9 = 11
108 ÷ 9 = 12

By 10

0 ÷ 10 = 0
10 ÷ 10 = 1
20 ÷ 10 = 2
30 ÷ 10 = 3
40 ÷ 10 = 4
50 ÷ 10 = 5
60 ÷ 10 = 6
70 ÷ 10 = 7
80 ÷ 10 = 8
90 ÷ 10 = 9
100 ÷ 10 = 10
110 ÷ 10 = 11
120 ÷ 10 = 12

By 11

0 ÷ 11 = 0
11 ÷ 11 = 1
22 ÷ 11 = 2
33 ÷ 11 = 3
44 ÷ 11 = 4
55 ÷ 11 = 5
66 ÷ 11 = 6
77 ÷ 11 = 7
88 ÷ 11 = 8
99 ÷ 11 = 9
110 ÷ 11 = 10
121 ÷ 11 = 11
132 ÷ 11 = 12

By 12

0 ÷ 12 = 0
12 ÷ 12 = 1
24 ÷ 12 = 2
36 ÷ 12 = 3
48 ÷ 12 = 4
60 ÷ 12 = 5
72 ÷ 12 = 6
84 ÷ 12 = 7
96 ÷ 12 = 8
108 ÷ 12 = 9
120 ÷ 12 = 10
132 ÷ 12 = 11
144 ÷ 12 = 12

$4 \div 2 \ =$ ____ $8 \div 2 =$ ____ $10 \div 2 =$ ____

$6 \div 2 =$ ____ $14 \div 2 =$ ____ $12 \div 2 =$ ____

$3 \div 3 =$ ____ $6 \div 3 =$ ____ $9 \div 3 =$ ____

$12 \div 3 =$ ____ $15 \div 3 =$ ____ $18 \div 3 =$ ____

$4 \div 4 =$ ____ $8 \div 4 =$ ____ $12 \div 4 =$ ____

$16 \div 4 =$ ⬚ $20 \div 4 =$ ⬚ $24 \div 4 =$ ⬚

$5 \div 5 =$ ⬚ $10 \div 5 =$ ⬚ $15 \div 5 =$ ⬚

$20 \div 5 =$ ⬚ $25 \div 5 =$ ⬚ $30 \div 5 =$ ⬚

$12 \div 6 =$ ⬚ $24 \div 6 =$ ⬚ $18 \div 6 =$ ⬚

$21 \div 7 =$ ⬚ $28 \div 7 =$ ⬚ $14 \div 7 =$ ⬚

~~Mark 3:24~~

If a kingdom is **divided** against itself, that kingdom cannot stand.

$2 \overline{)\ 48}$ $\qquad$ $2 \overline{)\ 32}$ $\qquad$ $2 \overline{)\ 20}$

$2 \overline{)\ 100}$ $\qquad$ $2 \overline{)\ 64}$ $\qquad$ $2 \overline{)\ 36}$

$5 \overline{)\ 250}$ $\qquad$ $5 \overline{)\ 40}$ $\qquad$ $5 \overline{)\ 45}$

$5 \overline{)\ 350}$ $\qquad$ $5 \overline{)\ 420}$ $\qquad$ $5 \overline{)\ 305}$

$3 \overline{)\ 36}$ $\qquad$ $3 \overline{)\ 333}$ $\qquad$ $3 \overline{)\ 627}$

3) 918   3) 303   3) 672

4) 350   4) 420   4) 848

4) 904   4) 480   4) 348

6) 36   6) 42   6) 48

6) 54   6) 12   6) 24

God created these animals too!

2 x 2 = [ ] 2 x 3 = [ ] 2 x 4 = [ ] 2 x 5 = [ ]

2 x 6 = [ ] 2 x 7 = [ ] 2 x 8 = [ ] 2 x 9 = [ ]

Write your own equation here:

~~**Genesis 1:29**~~

I have provided all kinds of fruit and grain for you to eat.

Arrays help with multiplication skills- complete each array and circle your favorite food.

___rows of ____ = ____   ___rows of ____ = ____

___ columns of ___= ___   ___ columns of ___= ___

__rows of __ = ___   __rows of __ = ___

__ columns of __= ___   ___columns of __= ___

The school bus has 12 rows of seats; there are 4 children in each row. How many children are on the bus? ____

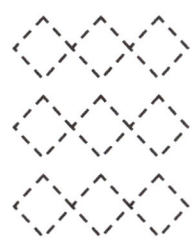

__rows of __ = ___   __rows of __ = ___   ___ rows of __ = ___

__ columns of __= ___   ___columns of __= ___   ___ columns of __= __

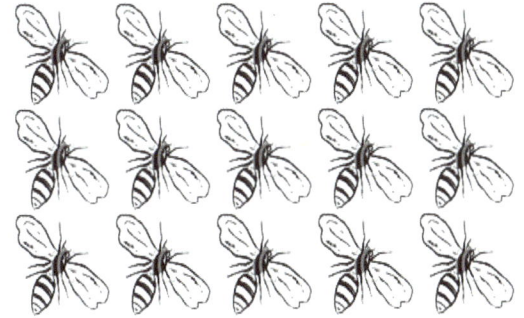

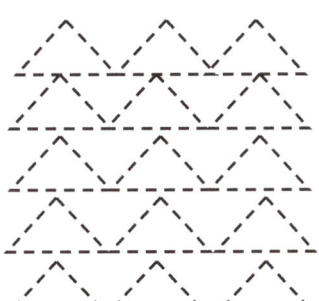

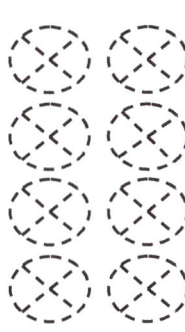

__rows of __ = ___   ___rows of __ = ___   ___rows of __ = ___

__ columns of __= ___   ___columns of __= ___   __columns of __= __

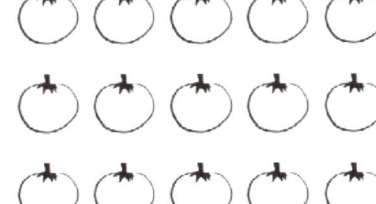

__rows of __ = ___   __rows of __ = ___   __rows of __ = ___

__ columns of __= ___   ___columns of __= ___   ___columns of __= ___

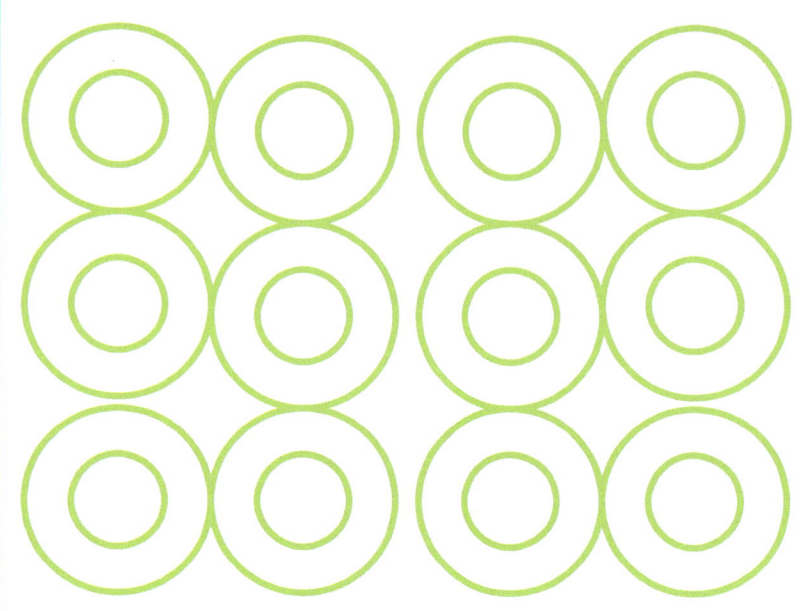

3 x 4 = _____

4 + 4 + 4 = _____

3 x 2 = _____

2 + 2 + 2 = _____

5 x 2 = _____

2 x 5 = _____

5 + 5 = _____

2 + 2 + 2 + 2 + 2 = _____

Multiplication Tables

2 x 1 = 2

2 x 2 = 4

2 x 3 = 6

2 x 4 = 8

2 x 5 = 10

2 x 6 = 12

2 x 7 = 14

2 x 8 = 16

2 x 9 = 18

2 x 10 = 20

3 x 1 = 3

3 x 2 = 6

3 x 3 = 9

3 x 4 = 12

3 x 5 = 15

3 x 6 = 18

3 x 7 = 21

3 x 8 = 24

3 x 9 = 27

3 x 10 = 30

4 x 1 = 4

4 x 2 = 8

4 x 3 = 12

4 x 4 = 16

4 x 5 = 20

4 x 6 = 24

4 x 7 = 28

4 x 8 = 32

4 x 9 = 36

4 x 10 = 40

Study these tables to help with your multiplication facts:

Do these:

2 x 4

=

3 x 6

=

4 x 8

=

2 x 9

=

Multiplication Tables

5 x 1 = 5

5 x 2 = 10

5 x 3 = 15

5 x 4 = 20

5 x 5 = 25

5 x 6 = 30

5 x 7 = 35

5 x 8 = 40

5 x 9 = 45

5 x 10 =50

6 x 1 = 6

6 x 2 = 12

6 x 3 = 18

6 x 4 = 24

6 x 5 = 30

6 x 6 = 36

6 x 7 = 42

6 x 8 = 48

6 x 9 = 54

6 x 10 =60

7 x 1 = 7

7 x 2 = 14

7 x 3 = 21

7 x 4 = 28

7 x 5 = 35

7 x 6 = 42

7 x 7 = 49

7 x 8 = 56

7 x 9 = 63

7 x 10 =70

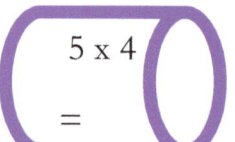

5 x 4

=

6 x 6

=

7 x 8

=

5 x 9

=

```
     15              15              21
x     5         x     4         x     2
```

```
     30              31              13
x     2         x     5         x     3
```

```
     15              11              12
x     3         x     4         x     4
```

```
     18              17              13
x     4         x     2         x     5
```

11 x 7 = ➡   13 x 7 = ➡   12 x 8 = ➡

27 x 3	13 x 4	14 x 2
10 x 3	10 x 1	10 x 4
10 x 5	30 x 3	23 x 3
44 x 2	22 x 4	5 x 5
26 x 5	15 x 5	5 x 4
7 x 2	4 x 3	16 x 4

Try these; see how many you can get correct!

$$\begin{array}{r} 16 \\ \times\ 5 \\ \hline \end{array} \qquad \begin{array}{r} 18 \\ \times\ 2 \\ \hline \end{array} \qquad \begin{array}{r} 9 \\ \times\ 2 \\ \hline \end{array}$$

$$\begin{array}{r} 7 \\ \times\ 5 \\ \hline \end{array} \qquad \begin{array}{r} 9 \\ \times\ 5 \\ \hline \end{array} \qquad \begin{array}{r} 3 \\ \times\ 3 \\ \hline \end{array}$$

$$\begin{array}{r} 18 \\ \times\ 3 \\ \hline \end{array} \qquad \begin{array}{r} 19 \\ \times\ 4 \\ \hline \end{array} \qquad \begin{array}{r} 5 \\ \times\ 4 \\ \hline \end{array}$$

God's blessings multiply every day!

$$\begin{array}{r} 81 \\ \times\ 4 \\ \hline \end{array} \qquad \begin{array}{r} 84 \\ \times\ 2 \\ \hline \end{array} \qquad \begin{array}{r} 47 \\ \times\ 7 \\ \hline \end{array}$$

$$\begin{array}{r} 50 \\ \times\ 3 \\ \hline \end{array} \qquad \begin{array}{r} 25 \\ \times\ 5 \\ \hline \end{array} \qquad \begin{array}{r} 19 \\ \times\ 3 \\ \hline \end{array}$$

$$\begin{array}{r} 31 \\ \times \quad 2 \\ \hline \end{array} \qquad \begin{array}{r} 40 \\ \times \quad 2 \\ \hline \end{array} \qquad \begin{array}{r} 45 \\ \times \quad 2 \\ \hline \end{array}$$

$$\begin{array}{r} 23 \\ \times \quad 3 \\ \hline \end{array} \qquad \begin{array}{r} 24 \\ \times \quad 2 \\ \hline \end{array} \qquad \begin{array}{r} 22 \\ \times \quad 4 \\ \hline \end{array}$$

$$\begin{array}{r} 21 \\ \times \quad 4 \\ \hline \end{array} \qquad \begin{array}{r} 35 \\ \times \quad 3 \\ \hline \end{array} \qquad \begin{array}{r} 41 \\ \times \quad 2 \\ \hline \end{array}$$

$$\begin{array}{r} 42 \\ \times \quad 2 \\ \hline \end{array} \qquad \begin{array}{r} 57 \\ \times \quad 2 \\ \hline \end{array} \qquad \begin{array}{r} 50 \\ \times \quad 2 \\ \hline \end{array}$$

$$\begin{array}{r} 32 \\ \times \quad 3 \\ \hline \end{array} \qquad \begin{array}{r} 42 \\ \times \quad 1 \\ \hline \end{array} \qquad \begin{array}{r} 55 \\ \times \quad 2 \\ \hline \end{array}$$

37 x 2	60 x 2	75 x 2
43 x 3	44 x 2	72 x 4
51 x 4	55 x 3	61 x 2
52 x 2	77 x 2	70 x 2
72 x 3	62 x 1	35 x 2

71 x 3	80 x 2	85 x 2
73 x 4	84 x 2	80 x 4
60 x 5	70 x 3	80 x 1
90 x 3	92 x 3	83 x 3
94 x 2	72 x 4	111 x 5
112 x 2	125 x 3	113 x 5

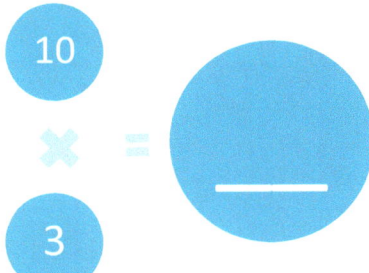

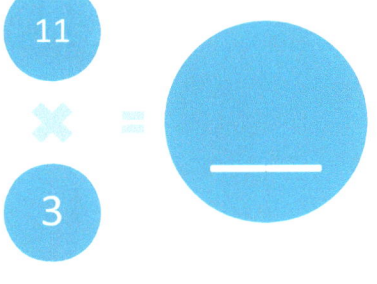

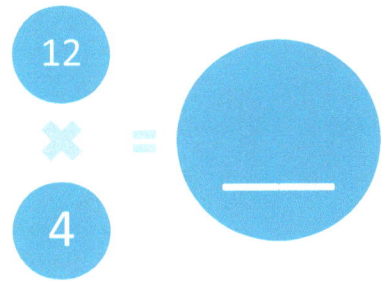

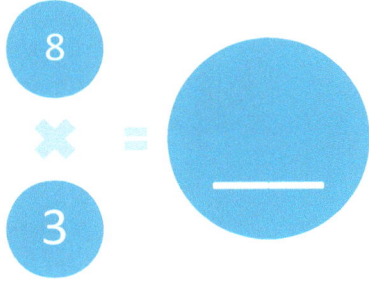

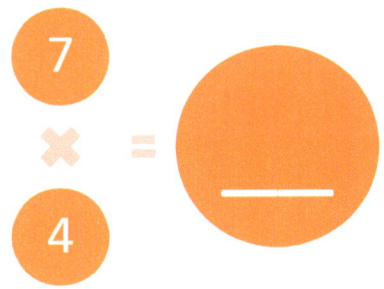

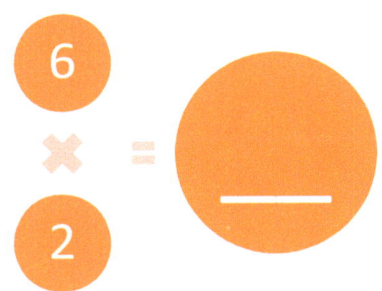

119 x 2	114 x 3	105 x 2
106 x 5	107 x 5	106 x 4
118 x 2	109 x 2	129 x 2
113 x 3	124 x 2	105 x 4
135 x 3	107 x 3	108 x 3

119 x 4	109 x 4	114 x 3
129 x 4	116 x 3	152 x 5
139 x 3	126 x 4	132 x 5

Free space, show your work here:

The outer wall of the side rooms was five cubits thick. The open **area** between the side rooms of the temple

Finding area helps with your multiplication skills! Area of a square is a^2

a

Find the area of the squares below:

11
11

6
6

7
7

12
12

10
10

9
9

2
2

5
5

8
8

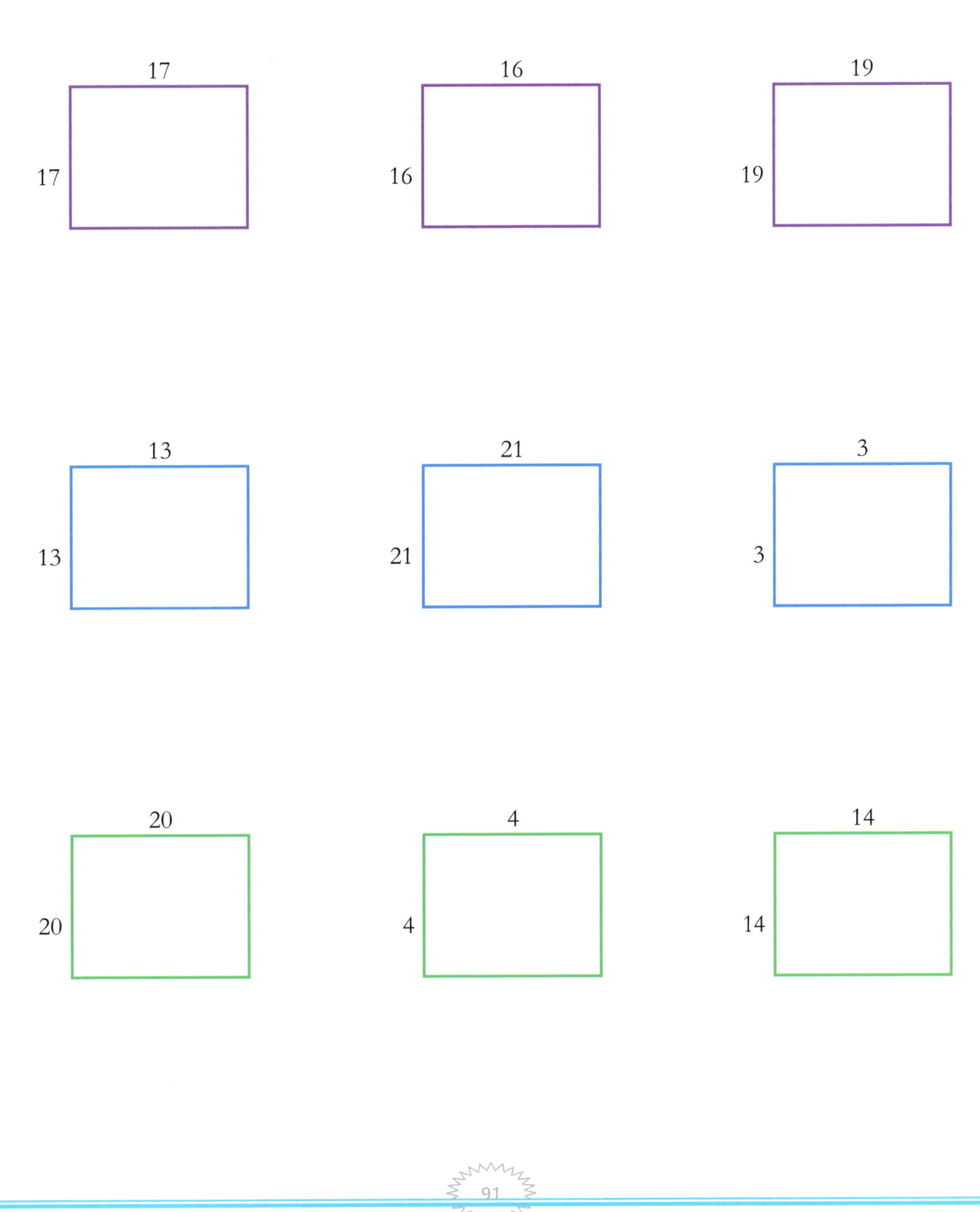

Area of a Triangle = ½ x <u>b</u>ase x <u>h</u>eight 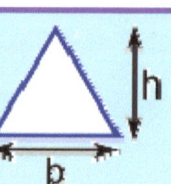 Find the Area-

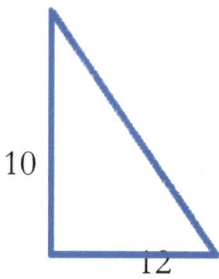

10   12

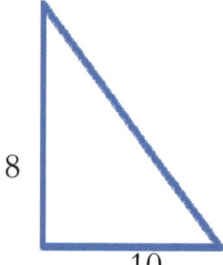

8   10

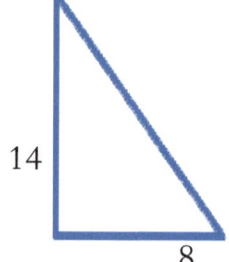

14   8

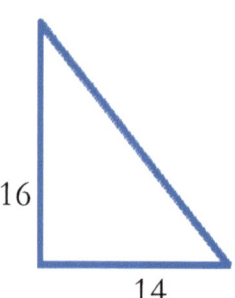

16   14

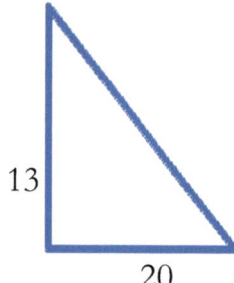

13   20

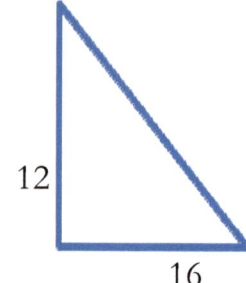

12   16

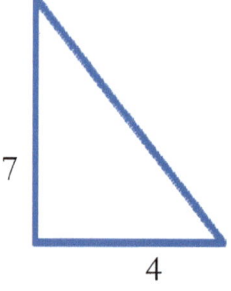

7   4

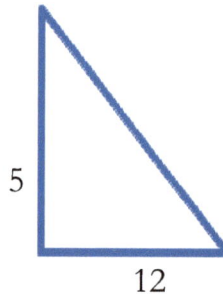

5   12

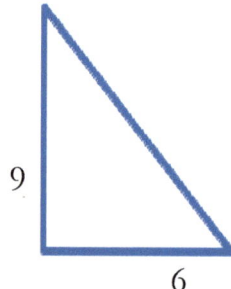

9   6

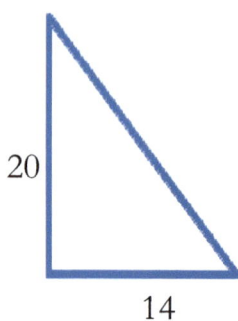

20   14

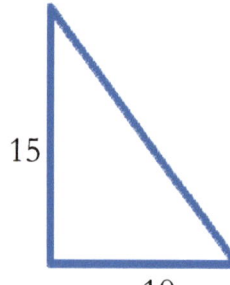

15   10

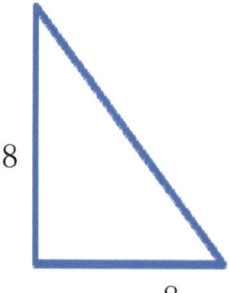

8   8

Area of a rectangle is w x h   Find the area of the rectangles-

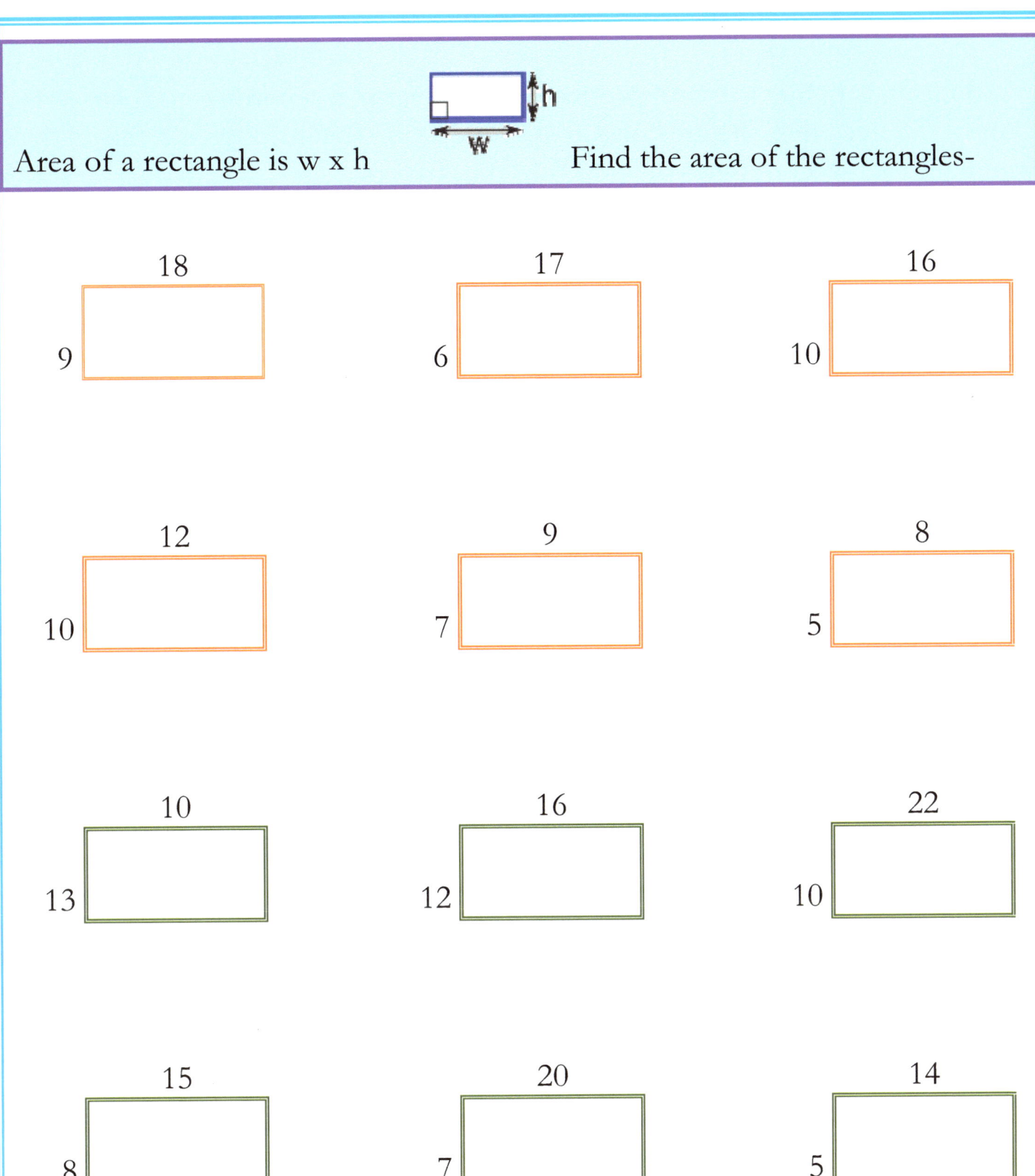

And God said, "Let the land produce living creatures according to their kinds: the livestock, the creatures that move along the ground, and the wild animals, each according to its kind." And it was so.

a) How many eyes do the sheeps have? _____

b) How many tails do the cows have? _____

c) How many feet do the lions have? _____

d) How many animals are there in all? _____

~~2 Kings 25:26~~

At this, all the people from the **least** to the **greatest**, together with the army officers, fled to Egypt for fear of the Babylonians

Arrange each set of numbers from the least to the greatest.

21   15   36   10

49   49   36   24

51   24   17   23

12   10   14   9

19   14   10   16

42   21   16   18

29   34   40   17

52   31   26   68

Arrange each set of numbers from the **greatest** to the **least**.

31   38   26   9

52   57   66   103

59   84   77   33

92   100   104   89

119   113   92   86

412   211   160   189

219   203   192   96

312   201   176   109

109   123   102   76

372   111   150   269

"Circle" the **greatest** number in each row and "underline" the **least** number.

1)　　400　　　　250　　　　390

2)　　640　　　　401　　　　320

3)　　550　　　　430　　　　110

4)　　89　　　　78　　　　52

5)　　125　　　　119　　　　201

6)　　190　　　　152　　　　178

7)　　175　　　　199　　　　207

8)　　360　　　　252　　　　176

9)　　521　　　　901　　　　102

10)　　179　　　　192　　　　108

> = Greater than   < = Less than

Write > or < on each line, example 500 > 450

a) 800 ____ 700

b) 1025 ____ 1159

c) 375 ____ 650

d) 230 ____ 175

e) 249 ____ 233

f) 1178 ____ 1039

g) 760 ____ 580

h) 1975 ____ 2015

i) 178 ____ 205

j) 765 ____ 539

k) 367 ____ 249

l) 982 ____ 716

m) 630 ____ 720

n) 775 ____ 890

o) 890 ____ 530

p) 579 ____ 321

q) 310 ____ 463

r) 665 ____ 412

s) 510 ____ 403

t) 555 ____ 212

u) 900 ____ 523

v) 745 ____ 232

w) 820 ____ 568

x) 905 ____ 619

~~~Congratulations, you did it!!~~~

~~Matthew 25:21~~

His master replied, '**Well done**, good and faithful servant! You have been faithful with a few things; I will put you in charge of many things. Come and share your master's happiness!'

WAY TO GO!

Let it be known that

....

did an awesome job

....

....

Signature   Date

www.ingramcontent.com/pod-product-compliance
Lightning Source LLC
Chambersburg PA
CBHW051020180526
45172CB00002B/422